Social Science Frontiers

Occasional Publications Reviewing New Fields
for Social Science Development

3

Social Aspects of Applied Human Genetics

by James R. Sorenson

Russell Sage Foundation
1971

Printed in the United States of America.
Library of Congress Catalog Card Number: 76-183973
Standard Book Number: 87154-819-4

Foreword

An earlier Frontier paper (No. 1), by Diana Crane, reviewed the links between the social and medical aspects of the prolongation and termination of life. Another area in which there are many problems at the interface of the social and bio-medical fields is genetics, including efforts at informing and counseling community members on the consequences of their reproductive behavior and in modifying their conceptive practices. As part of the Foundation's program on the social adaption of man to technological change in medicine and innovation in medical practice, it is supporting work on genetic counseling by Professor Sorenson.

This Frontier paper comes from a review of social service and bio-medical literature that provides a social perspective on human genetics and his own research endeavors. In this document, Professor Sorenson identifies a number of needed research areas and issues for investigation. It is hoped its publication will stimulate more work in the area. At the present time a wide range of research opportunities exist, studies that can contribute to social science knowledge and which may be of value for policy making and program development. The Foundation will be pleased to advise and collaborate with interested parties.

Howard E. Freeman
Russell Sage Foundation

Contents

Social Aspects of Applied Genetics

Recent advances in science and medicine are increasing man's control of the quantity and quality of human populations. The development of highly effective and relatively efficient birth control techniques permits expanded regulation of population size and growth. In addition, advances in medical genetics are making possible growing intervention and manipulation of the genetic quality of human populations.

Progress in medical genetics can alter man's role in evolution. Man is no longer limited to passive acceptance of all inherited characteristics but is rapidly expanding his technological capacity to include the active treatment, selection, and elimination of many individual genetic attributes. These developments pose complex questions of a moral, ethical, political, psychological, or economic nature. For instance, what genetic attributes or constitutions are desirable? Who is to decide? Should genetic anomalies be reduced in a population? Who shall say how or when? As with most technological developments, knowledge that permits increasing intervention and control of the genetic quality of life is accumulating more rapidly than is man's ability to apply this knowledge wisely.

Some of the technological developments that permit control of the quality of human life have not precipitated serious problems. In most Western societies medical research and practice have achieved near complete control of many of the major infectious diseases that have plagued mankind for centuries; such control has met little resistance. A case in point is the development of polio vaccine and the subsequent virtual elimination of poliomyelitis.

The success of such medical advances was dependent on several things, especially prevailing values. For example, the implementation of programs to control infectious disease required a value system in which disease was interpreted as a natural event. If parents felt shame or guilt for the infectious diseases suffered by their children, attempts to treat the disorders and to develop curative methods would have been obstructed. Also, such a value system had to provide for the approval of man's active intervention to control disease. The doctor in Western society has not only received approval for his intervention in disease control but also has achieved a great deal of esteem and prestige from the society he has served.

The idea of treatment and intervention in man's genetic health is not universally accepted today, by either the general public or the medical profession. This is due in part to existing values and beliefs. Parents often experience guilt and shame for genetic disorders in their offspring (Lynch, et al., 1965). More important perhaps, many people including some medical personnel, believe that medical procedures are not capable of correcting or treating any genetic disorders, or that doctors and parents should not attempt genetic intervention (Lynch, 1969).

Through the ages man has interpreted the significance of genetic anomalies in many ways. Sometimes, anomalies were interpreted as the favor of the gods, and at other times they were considered to be portents of divine wrath (Reisman and Matheny, 1969). Today where these beliefs still linger, they limit the use of genetic knowledge by the public and by medical practitioners. Counterbalancing these beliefs and values is a rapidly increasing technological capacity to treat and to select certain genetic conditions. The eventual role of applied human genetics in medical practice and society will reflect the complex intertwining of existing values and beliefs with increasing technical capacity. Neither factor alone will predict man's orientation toward an intervention in his genetic future.

Recent Developments in Clinical Genetics

One of the earliest applications of genetic knowledge in medical practice was genetic counseling, also referred to as "inheritance counseling." This is a form of medical service through which people gain information about their genetic constitution or that of their children. Genetic counseling has had a short history. For much of this history doctors were limited to informing families that they had a disease or disorder that "ran in the family." Such medical counsel was of little practical use to families. With the advent of Mendelian genetics and much later the linkage of selected diseases with specific modes of inheritance, genetic counseling became more precise. Doctors could give parents with an identifiable genetic disorder statements of the risk that any child they had would exhibit the genetic defect. For example, a couple might have had a child who died from cystic fibrosis, a lethal genetic disease of childhood. A counselor making this diagnosis would inform the couple that they faced a risk of approximately 1 in 4 of having another child with cystic fibrosis (Carter, 1969). The couple could then use this information to make a decision about future reproductive behavior.

Genetic counseling, as practiced until recently, was limited primarily to the calculation and delivery of such risk statements. Prospective parents could evaluate the risk in a given case and then act on their decision. If they chose to take a risk, they had to live with the child whether he was normal or not.

Several recent advances are changing the nature of genetic counseling. The development of amniocentesis, a procedure by which fetal chromosomes are analyzed for abnormalities, permits intra-uterine detection of several genetic and most major chromosomal defects (*Fogarty International Center Proceedings*, 1970). Using this technique, parents may choose to abort a fetus with an identifiable defect and to have only healthy children. For example, Down's Syndrome, more commonly known as mongolism, a congenital moderate-to-severe form of mental retardation due to specific chromosomal abnormalities, can be detected in utero with almost total accuracy and little apparent risk to the mother or fetus (*Fogarty International Center Proceedings*, 1970). With this procedure the mother need no longer carry such a fetus to term, if the parents so decide and if their doctor and the state concur. The risk of having such a child can thus be removed. Recent developments also permit the detection of an increasing number of genetic defects in persons who carry a genetic anomaly but do not clinically exhibit the disease. For example, carriers of sickle-cell anemia, a form of anemia due to an abnormal type of hemoglobin in red blood cells, limited almost entirely to black populations, can be identified (Carter, 1969). This procedure permits afflicted couples to be advised of their condition prior to having children. Thus, genetic counseling can be given before the birth of any afflicted children, rather than after, as in the past.

In addition to an expanded technical base, many other factors have increased the potential application of genetics in medical practice. First, the role genetic endowment plays is being delineated in more and more diseases. It apparently plays a minor role in the etiology of many diseases, such as cancer and heart trouble; a moderate role in others, such as diabetes; and a major, if not determinate, role in a third class of disease, such as cystic fibrosis, sickle-cell anemia, and Down's Syndrome (Carter, 1969). Until now, genetic counseling has been limited primarily to diseases falling in the third class, but it can be used to advise individuals who have diseases of the first two classes. In these situations clients can be informed of the risk of occurrence of the disease and can be advised to take precautionary measures, such as dietary and environmental management, to reduce their risk of disease. In addition, increasing research efforts and discoveries in behavioral genetics, while not yet applicable, indicate that genetic counseling might eventually involve the role that inheritance plays in the appearance of selected aspects of psychological functioning and possibly some aspects of social behavior (Lindzey, et al., 1971).

Of the diseases in the population that are major threats as health problems, those with a genetic base are constituting an increasingly larger proportion (Carter, 1969). In controlling infectious diseases man has found that imperfections or peculiarities in his own genetic constitution are becoming health threats. This factor will certainly operate to increase the demand for early and extensive appilcation of medical genetics to human populations by both professional medical groups and the public.

Changes in public attitudes and practices are increasing public demand also for more extensive programs in medical genetics. With the increasingly widespread acceptance of population control, it seems reasonable that parental concern for the health of their smaller families will increase. This could include concern for the genetic health of their children. In addition, the acceptance of birth control practices reflects new values and changed attitudes toward reproductive behavior. Not only can reproduction be planned, but with the new developments mentioned above the health of many children who risked disease can now be reasonably assured.

Finally, there may be considerable economic advantages in developing mass screening programs of people likely to pass on specific genetic diseases to their offspring (Scriver, 1970). Therapeutic abortion of fetuses afflicted with Down's Syndrome or other severely disabling diseases detectable in utero can save families much sorrow, and save both the state and the families from a severe financial burden (Danes, 1970). The development of intra-uterine detection techniques makes such screening programs feasible. The changing moral and legal climate surrounding abortion suggests that the number of therapeutic abortions will increase in the near future.

Increased application of genetics in medicine will not necessarily be automatic, however. The use of genetic knowledge in medical practice has given rise to many controversial issues in the recent past and continues to do so. How large a risk should prospective parents take? Should parents make the decision? Should abortion of a fetus be permitted on the grounds that it will be abnormal? Should carriers of severe genetic defects be forbidden to marry, or to have children? These questions are in every respect—ethically, morally, politically, emotionally, legally—difficult to answer. More questions will arise as knowledge of human heredity advances.

The increasing intervention in and control of genetic disease and the selection or avoidance of certain genetic constitutions have fostered debates within and between many different groups in society. Medical professionals, life scientists, lawyers, ministers, biologists, and philosophers are engaged in discussions concerning the proper use of such knowledge. A prerequisite for intelligent discussion of these issues is information about how genetic knowledge is used today. Which doctors give genetic counseling? Which parents seek counseling? What types of reproductive decisions do people make when faced with genetic disorders? Very little information is available on these topics. If man is to understand and to employ the potential good inherent in medical genetic advances, he must begin by determining by whom and how these advances are used. The social sciences, by providing such information, can make significant contributions to discussions on the use of genetic knowledge.

Human Genetics in Medical Practice: A Review and Analysis

Because the application of human genetics in medical practice is relatively new, few studies explore the psychological and sociological aspects of genetics in medicine. The existing literature is both meager and scattered.

There are numerous medical publications concerning the clinical nature of genetic counseling, which is the major use of medical genetics today (Sorenson, 1971a). This literature includes discussions of problems that genetic counselors face, and the types of services they provide. A small amount of literature reports the general public's knowledge of specific genetic disorders. In addition, there are a few studies of the reproductive decisions made by couples after genetic counseling. There are no detailed studies of the distribution in the United States of medical genetic facilities or their services. Finally, little mention is made of the economic aspects of medical genetics.

Most of the available information has been reported by individual medical geneticists or counselors and, being based on their personal experiences, is of limited use in generalizing about medical genetics. There is thus considerable need for extensive social science research on applied human genetics.

Public Knowledge and Use of Medical Genetics

Extensive sociological literature suggests that knowledge about certain kinds of infectious diseases and action taken to alleviate those diseases is inversely related to social class (Feldman, 1966; Mechanic, 1968). The assumption that knowledge about various genetic disorders would vary according to social class is substantiated by existing literature.

In a national survey Feldman (1966) found considerable variation in public knowledge about various diseases. For example, whereas nearly 70 percent of his sample knew at least one correct symptom of polio, only 48 percent could correctly name a symptom of diabetes, a genetically based metabolic disorder (Feldman, 1966). More specifically, knowledge about diabetes was positively associated with education and income, established indicators of social class. The experiences of many genetic counselors support this observation. Their practice suggests that in the past, and to a large extent today, there has been little utilization of medical genetic facilities by the poor (Juberg, 1966).

The fact that historically the majority of people receiving genetic counseling have been of the middle or upper social classes is a reflection of several factors. First, the middle class, and apparently their doctors as well, have been more informed than other classes about the availability of medical genetic services. Second, since the use of genetic counseling has in the past depended largely on self-referral, the willingness of the middle class and the reluctance of the lower classes to seek needed medical assistance have been important (Mechanic, 1968). Economic constraints, while of some importances in shaping the use of medical services, are not always of primary concern (Myers and Schaffer, 1954). Until recently, most genetic counseling was free, with minimal charges for necessary laboratory work. The factors determining the distribution of services appear to have been essentially special knowledge of the availability of such services, plus sets of values about and attitudes toward the procreative process that legitimated intervention and control.

The social distribution of health information is itself determined in part by the social sources of that information, that is, the process of knowledge dissemination. Feldman (1966) suggests that for diseases such as cancer and polio, essentially nongenetic diseases, a primary source of information is the mass media. His research, as well as several additional studies (Koos, 1954; Deshaies, 1962), suggests variation across social classes, however, in the relative importance of various information sources for knowledge of infectious disease. Generally, the mass media are basic information sources for the more educated segments of the population, while personal contacts serve as primary information sources for the less educated segments (Feldman, 1966). Doctors play a minor role in educating the public. Their role is apparently limited to informing the patient about the particular disease he has (Feldman, 1966).

Public knowledge about genetic disorders does not appear to be disseminated in the same way as knowledge about nongenetic disorders, especially for the more educated segments of the population. Feldman reported that the most important source of information about diabetes was an afflicted relative or close friend. This suggests that general knowledge about genetically based diseases depends more upon interpersonal channels of communication than does knowledge about infectious diseases, and that individuals are likely to be acquainted with such diseases only if a family member or close friend is afflicted.

The social distribution of knowledge about genetic disease reflects the fact that genetic considerations are of limited interest to most people. Concern for these disorders is generally confined to those afflicted, or to those in their reproductive years who have family or friends suffering from genetic disorders. Experiences of counselors indicate that genetics is of minor concern for most people when they select marriage partners. Requests for premarital counseling constitute a small segment of a counselor's time (Reed, 1963). In those cases the most frequent questions concern consanguinity, not the existence of a genetic problem in one partner or the other. The most frequent requests for counseling come from the parents of a genetically defective child (Reed, 1963). It is at this point that the parents become concerned about genetic health and seek counseling to avoid further defective children or simply to relieve anxiety. Counselors are also often contacted for advice by state and local agencies such as state institutions and adoption agencies. Occasionally, counselors are contacted for assistance by the courts in a paternity suit (Reed, 1963).

Taken together, these observations suggest that in the past and at present medical genetic services are not used by all who could benefit from them, nor are they used at the most opportune time. While medical genetics is inherently a preventive form of medicine, it seems to be used in response to crisis situations—after genetic problems occur, rather than before.

The Nature and Scope of Genetic Counseling

There are many forms of genetic counseling. The people who practice genetic counseling have various educational backgrounds: some are research scientists, others medical doctors (Sorenson, 1971b). Genetic counseling is not yet a medical specialty. Because most physicians lack training in human genetics, counseling is usually limited to those who have specialized knowledge about human genetics, although they may or may not be medical doctors.

In the late 1940's and early 1950's there were perhaps 10 or 12 genetic counselors in the United States (Reed, 1963). Today there are about 200 counseling units (Lynch and Bergsma, 1971). With mounting discoveries and an increasing awareness of the need for counseling, this impressive growth should continue in the near future.

The most common task facing the genetic counselor is the determination of the risk of recurrence of a genetic defect in a family (Reed, 1963). Supported by Mendelian genetics, the notions of dominant and recessive autosomal or x-linked modes of inheritance, the counselor can often provide specific information about recurrence risks (Lynch, 1969). Although many genetic disorders may follow the classical inheritance modes of Mendelian ratios, many do not. In these cases the counselor must rely upon empirical statistics of risk. These statistics have been derived from studying the incidence of a disease in many families (Neel, 1958). From these observatons the counselor can provide the clients with approximations of the recurrence risk of a specific disease. When the counselor is equipped with neither Mendelian ratios nor empirical estimates of risk, he can inform his clients only of his ignorance.

In addition to statistics of risk the counselor today can employ a battery of new techniques that permit increased detection of genetic and chromosomal abnormalities in utero. These techniques include amniocentesis, examination of the genetic and chromosomal health of the fetus, and fetoscopy, a technique that enables visual inspection of the fetus (*Fogarty International Center Proceedings*, 1970). Advances such as these are applicable only when a chromosomal or genetic defect has a structural or metabolic effect upon the developing fetus. Many genetic defects cannot yet be detected by these techniques. While amniocentesis appears to be used primarily for the detection of Down's Syndrome, fetoscopy is still in a largely experimental stage. These techniques are relatively expensive and not widely available.

The Role of the Genetic Counselor

Genetic counseling as a form of medical service differs from traditional medical practice. A complex relationship exists between the counselor and the counselee in genetic counseling. While some counselors perceive the relationship in the traditional doctor-patient manner, others perceive it as a counselor-client relationship (Sorenson, 1971b). Because the rights and obligations of both parties of this relationship have not been completely defined, the practice of counseling is probably based primarily on the beliefs, attitudes, and skills of the individual counselor.

Some counselors feel their primary task is simply to inform the counselees of the risks involved. They believe that to try to influence the decisions of the counselees is to go beyond the professional responsibility of the counselor. While a counselor may feel that the risk involved in a specific case is great or that a given disorder should not be transmitted to new carriers, he should refrain from telling the counselees what they should do. Often, in this situation, the counselor will define his relationship with the counselees as a counselor-client relationship (Reed, 1963). In so doing he is stressing the learning nature of the relationship. The client is there to be informed, not to be healed.

Characteristic of the traditional doctor-patient role is the healing relationship in which the counselee yields discretionary power to the doctor (Bloom, 1963). Some counselors describe their relationship with their counselees in this fashion (Lynch, 1969). The counselor may suggest how patients should use the information he gives them, and his concern for the patient goes beyond a mere statement of statistics of risk. For example, the counselor may discuss in detail the psychological consequences of an abnormal child, or the economic burden of genetic abnormality. He may also point out to the parents the impact on the family of a defective child. Finally, he may discuss with them the problem of dealing with an afflicted dependent when they advance in age and retire (Fletcher, 1971). All of these are important factors in conveying the meaning of genetic risks to parents and undoubtedly have an impact on the parents' decisions about future reproductive behavior.

Because his role falls within the purview of medical service, the counselor is probably approached by most people in the traditional patient manner. This orientation requires the counselor to assume much more discretionary power than most want. The meaning of a risk is complex, as is the meaning of disease or abnormality. While many counselors and counselees may agree that it is good to avoid having mongoloid children, there is less agreement as one moves from physiological abnormalities to such problems as an XYY chromosome anomaly and the hypothesized link with social pathological behavior (Public Health Service, 1970).

The information the counselor gives, as well as the way in which he gives it, is crucial in defining the situation for parents. For example, if the counselor indicates that in the United States about 1 in 50 births will result in a severely abnormal child (Carter, 1969), he can label the risk high or low. If he says that the odds for a *normal* birth are 3 in 4, or conversely the odds for an *abnormal* birth are 1 in 4, the facts are the same but the impact of the information on the parents is likely to be reversed. Because counselors are generally aware of the problems of giving genetic counsel, many are concerned about exerting too much influence on the parents.

The Delivery of Genetic Counseling

The practice of genetic counseling appears to be moving from academic departments to departments in medical schools and in public health centers (Sorenson, 1971b). Today, the typical counselor is more often an M.D. or an M.D.-Ph.D. than the traditional Ph.D. geneticist of the past. These shifts have numerous implications for both the delivery and the scope of genetic counseling.

Whether counseling facilities are located in medical schools or public health clinics makes a difference in the distribution and utilization of counseling. Most doctors in medical schools occupy the dual role of teacher and researcher. Research interests are normally pursued by intensive study of a particular facet of disease. For example, if a doctor-researcher interested in genetics focuses his attention on cystic fibrosis, he will become professionally identified through publications and more informal channels as a specialist in cystic fibrosis. As a consequence, both the dictates of his research and his professional identification will restrict the types of patients he counsels. In addition, if the medical school in which the researcher works does not have a large department or program in human genetics, as is often the case, it will become identified as specializing in counseling on limited genetic topics. Patients with other genetic problems will not be referred to it and must go elsewhere for counseling.

Genetic counseling is beginning to be practiced in some public health centers (*Symposium on Human Genetics in Public Health*, 1964). With this development, the proportion of lower socio-economic groups receiving counseling will most likely increase. Public health nurses, whose professional duties are primarily concerned with lower-income groups, will become a major force in the dissemination and application of genetic counseling. The public health nurse can provide the family with attention at home, a service seldom available in genetic counseling today (*Symposium on Human Genetics in Public Health*, 1964). Genetic problems can, and often do, have a profound impact upon the social-psychological stability of the family (Zuk, 1959). Family follow-ups and counseling should provide for more effective genetic guidance.

Reproductive Decisions After Counseling

There is limited information on reproductive decisions made by patients after counseling. Certainly many factors are important in determining these decisions: (1) the size of the risk, (2) the severity of the potential abnormality, (3) the social and private attitudes of the parents toward abnormality, (4) the economic capacity of the family to endure the burden of a genetic disease, (5) the genetic health of existing childern, and (6) the type of counseling parents receive. In a recent study, Carter (1966) presented a follow-up of the reproductive behavior of 169 couples referred to a genetic counseling clinic in England. His study revealed that the magnitude of the risk had a large impact on the future reproductive behavior of the couples. Fully two-thirds of the high-risk parents, as against one-fourth of the low-risk group, did not have additional children. (Of course, factors other than the magnitude of the risk affect these reproductive decisions.)

As we have seen, couples who are faced with the prospect of a high risk of a serious defect may decide to take the risk and have a baby. If they decide not to take the risk, however, several options are available to them: (1) they may give up plans for having additional children; (2) depending on the nature of the genetic disorder, they may decide on artificial insemination from a donor; or (3) they may decide to adopt a child. As with the decision whether or not to take a risk, decisions among alternative ways to have a child are influenced by many factors. Unfortunately, there are no published data either on these decisions or on the effects upon the decisions of social class, religious beliefs, legal constraints, and financial conditions. Carter (1966) does indicate that within his sample of high-risk couples who decided not to have any additional children adoption was low. In several cases the wife was sterilized, and in many cases the husband.

With few exceptions, much of the preceding information is based on the personal experiences of genetic counselors. While this information is important, more systematically gathered knowledge is needed. The increasingly important role that genetics will play in medicine in the future requires that we understand how people react to and use genetic knowledge, whether they are professional medical personnel or the public. In addition, the important social, ethical, and legal problems beginning to arise from the development of new medical ideas and technology require solid knowledge of the extent and use of genetic counseling today. Knowledge is being applied and precedents are being set that will shape the use of future discoveries.

Areas of Needed Research

Because psychological and sociological research on applied human genetics has been very limited, extensive studies are needed in at least five major areas: (1) the social organization of medical genetics; (2) the training, practices, and attitudes of genetic counselors; (3) the clients of genetic counseling, their problems, attitudes, and decisions; (4) the legal implications and social consequences of applied human genetics; and (5) the economic aspects of the delivery and utilization of medical genetics.

The Social Organization of Medical Genetics

Little is known about the social organization of medical genetics. The impact on counseling of differences in professional composition of clinics needs to be investigated, as does the effect of the organization of medical genetics on utilization patterns.

In the absence of exact historical documentation on the use of genetic knowledge in medicine, socio-historical analysis is an excellent method to study the progressive transition of genetic knowledge from its purely scientific orientation to its increasing inclusion in the domain of the medical profession. Prerogatives in the application of this knowledge are now being claimed by medical groups, as evidenced by an increasing array of articles in medical journals outlining the rights and obligations of the genetic counselor.

Particularly informative would be a socio-historical case study of medical counseling for Down's Syndrome, or mongolism. Down's Syndrome has very serious consequences for the afflicted person and his family. The family must adapt to a mentally deficient child and assume a tremendous economic burden if institutionalization is required (Evans, 1954). Counseling for this abnormality has progressed from the stage where counselors could only recite figures for recurrence risks to the current stage of active intervention in the control of the disorder through amniocentesis and therapeutic abortion. As a result of these developments, the control of Down's Syndrome has clashed with both societal values and legal prohibitions, especially with regard to abortion. As more genetic disorders come under this type of control, similar conflicts will probably arise. A study of the problems and issues in this case could provide considerable information on the probable course of other developments in medical genetics.

Surveys are needed of the current distribution of medical genetics and the kinds of medical services it offers. Programs offered by large national groups such as the National Foundation—March of Dimes and the National Genetics Foundation need to be studied. These organizations play an important role in informing the medical profession and the public about medical genetics and in bringing about changes in values and attitudes toward genetic defects.

Effective utilization of scientific discoveries often requires changes in the social system of medicine as well as acceptance by the public. The medical profession historically has been prepared to diagnose and treat nongenetic diseases. Despite the increasing need for medical genetics, current medical training does not include extensive study of genetics, nor does the nature of this training provide for effective utilization of existing genetic knowledge (World Health Organization, 1962). Studies of the training of doctors in human genetics are needed. Surveys of the attitudes and opinions of doctors concerning genetics and genetic counseling could provide insight into the medical profession's acceptance of advances in genetics.

The acceptance of genetic counseling in everyday practice poses numerous problems for doctors. For example, traditionally doctors have been trained to treat a large number of people who suffer from a relatively small number of major diseases. The application of human genetics in medicine reverses this procedure. In genetics there is a large number of diseases but a relatively small number of people who suffer from each disorder (Danes, 1970). Various facilities such as national diagnostic centers and networks are being developed to deal with this new situation (Danes, 1970). Studies are needed concerning the effectiveness of these developments and their implications for the integration of medical genetics into general medical practice.

Also important in the acceptance of genetic counseling by physicians will be their willingness to perform a type of medical counseling in which the traditional role of the doctor as healer is to a large extent not yet possible. Freidson (1970) has recently noted that the basic focus of the practice of medicine is the application, not the creation, of knowledge: "[Medical practice's] focus is on the practical solution of concrete problems, [and] it is obliged to carry on even when it lacks a scientific foundation for its activities: it is oriented toward intervention. . . ." To a large extent, the application of human genetics in medical practice, with the exception of intra-uterine analysis, does not yet permit much practical intervention by doctors into the health status of their patients. This lack of practical intervention has probably delayed the acceptance of human genetics into medical practice, even though human genetics can provide parents with important information about the risks of producing a genetically defective child. Studies of the attitudes of practicing physicians concerning human genetics and its perceived role in medicine could begin to clarify the complex connections between the degree to which scientific knowledge permits intervention or action by a doctor and the willingness of the doctor to incorporate the knowledge into his everyday practice.

In addition to the practical characteristics of new knowledge, the social and psychological characteristics of practicing physicians shape patterns of acceptance. Numerous studies have charted the social and psychological correlates of acceptance of various types of innovations by physicians (Menzel and Katz, 1955-56). Studies such as these are needed to find out which doctors in what situations are incorporating medical genetics into their practice. Are the innovators largely young doctors? How do the private opinions and beliefs of a doctor about eugenics, abortion, and religion affect his acceptance and use of medical genetics?

It would be particularly important to study the training of gynecologists, obstetricians, and pediatricians. These doctors occupy strategic monotoring positions in the reproductive cycle and thus assume a primary role in linking genetic counseling services with potential clients. Research could reveal the extent to which the fetus is gradually being seen by medical practitioners as a patient, and the implications of this transition for medical practice.

In order to save parents the psychological and economic burden of raising an abnormal child, obstetricians often permit newly born defective children to die. The conditions under which this type of decision is made merit study. How does the severity of the anomaly affect the doctor's decision? Is the economic condition of the family a factor in the decision, and does it make a difference whether the anomaly is physical or mental in nature (Crane, 1969)?

Finally, international comparative studies are needed on the social organization of medical genetics. Several Scandinavian countries, including Sweden and Denmark, have a long history of applying genetic knowledge in medical practice. Other countries have state-regulated programs in genetics. In certain sections of Italy, where the genetic disorder *thalassemia*, a debilitating anemia, is widespread, premarital counseling is required by the state, although decisions about reproduction are left to individual couples (Carter, 1969). Several investigators have studied intensively public acceptance of various medical advances, including polio vaccination programs (Clausen, 1954; Deasy, 1956). These studies generally reveal that acceptance of medical innovations is associated with education and exposure to certain forms of mass media. Similar studies are needed on public acceptance and use of medical genetics nationally and internationally. Are there significant differences among cultures in the acceptance of medical genetics? Do the nature and structure of the family unit as they vary culturally affect the use of genetic counseling?

A particularly interesting comparative analysis would include an examination of medical genetics in Russia and in the United States. A major basis for knowledge concerning genetic counseling in the Western world has been Mendelian laws of inheritance and the theory of inherited characteristics. Although Lysenko's genetic theories are presently discredited, they undoubtedly have influenced the use of genetics in Russian medical practice. Research on the extent and use of genetic counseling in Russia historically, as well as currently, could reveal the impact of ideological values on medical practice.

Taken together, the studies outlined above will provide important information about the extent, use, and social organization of medical genetics on an international as well as a national basis. Such research will be of considerable value in understanding the problems and issues involved in applying scientific genetic knowledge to human populations.

Genetic Counselors

A topic of utmost concern in the study of genetic counseling is the professional element of the counseling session—the genetic counselor. Training, research interests, and orientation to his client all contribute to the effectiveness and impact of the counselor's efforts.

Although most medical schools offer only elementary courses in medical genetics, general practitioners engage in some genetic counseling. Counseling today, however, is probably limited mainly to personnel with special training in genetics. The extent of counseling offered by the family doctor or by the trained specialist has not yet been studied. Research is needed to specify the conditions under which family doctors do counseling and under which they refer patients to specialized counselors. Research is also needed on the training of counselors and the impact this training has on their counseling. As discussed, some counselors have only academic backgrounds, while others have been trained as medical doctors. There could be great variation and interest in the patient and in the capacity to diagnose diseases and specify their medical implications.

Many people have recently expressed concern about the dehumanization of medical service because of its increased scientific base and delivery by scientifically trained personnel (Wilson, 1966). Scientific orientation need not always adversely affect the doctor's concern for the patient, however. For example, in a recent study of medical educators in medical schools in the United States, Babbie (1970) found that the degree of scientific orientation of a doctor did not significantly affect his concern with the care and treatment of his patients. Babbie's study was conducted in hospitals, where role prescriptions that define concern for the patient are salient. What is the situation, however, when a doctor interacts with patients outside the hospital in his private office, with fewer constraints and much less surveillance, as often occurs in genetic counseling? Does the doctor's degree of scientific orientation have an impact in these situations? Research similar to Babbie's but conducted in less structured medical settings is needed before we can determine the full relationship between a doctor's scientific orientation and his treatment of patients, and how this affects medical genetics.

Many counselors are located in medical schools where research and teaching are as important as treatment. We would expect that the research interests of these doctors might have an impact on the nature and scope of their counseling. Do they accept clients only on the basis of their research interests? If so, do they perceive their clients as people to be studied rather than as people to be counseled? Again, studies similar to Babbie's (1970) could begin to answer some of these questions.

Finally, we need careful study of the types of counsel given by genetic counselors and how counselors perceive their role. Existing medical literature suggests that the normative structure surrounding genetic counseling is far from established. Studies similar to Szasz and Hollender's (1956) work on the orientations of the doctors to patients are needed, focusing on changes in the nature of medical practice brought about by medical genetics. Doctors vary significantly in their orientation to patients. Some are much concerned with the psychological aspects of the doctor-patient role, which they consider to be a psychotherapeutic situation (Lynch, 1969). Others are authoritarian and see no need to concern themselves with the psychological aspects of the role. To what degree do such differences exist among genetic counselors, and how do they affect counseling? To what extent do genetic counselors discuss with patients the special significance of a given disorder? Do they discuss the medical complications of a disease? Such information is very likely to have an important effect on patients' decisions. Research could also determine if counselors systematically label risks of a certain magnitude as high or low, and the terminology counselors use to discuss diseases.

Clients of Genetic Counseling

Studies are needed of clients of genetic counseling and the referral systems that link client with counselor. Further study is required to examine clients' decisions and how the counseling situation affects reproductive decisions and strategies.

Because most counselors report today that a significant proportion of their referrals are self-generated, surveys are needed on the public's attitudes and beliefs about genetic defects and their knowledge of available services. Feldman's work (1966) suggests that the informational basis of the public's knowledge about genetic disorders is limited primarily to interpersonal sources. Up to this point mass media efforts have been extremely limited but they do have an impact. Counselors report that variations in demands for counseling follow specific patterns. Whenever there is a story in the popular press concerning medical genetics, counselors are flooded with requests for their services. However, this demand soon subsides until another article is published. This suggests that while the need and the desire for genetic counseling are widespread, most people are probably unaware of the possibilities of such services. Most physicians probably do not inform their patients of the availability of medical genetic services, if, indeed, the physicians themselves are aware of the services. Research on public knowledge similar to Feldman's study but focused more specifically on genetic disease could reveal the potential demand for medical genetic services and the problems that might be encountered in developing mass genetic screening programs or policies that require counseling prior to marriage or child bearing.

The application of genetic knowledge to human populations by means of screening programs can become very complex when genetic disorders are limited to specific segments of a population. For example, the genetic disorder of sickle cell anemia is found to be concentrated in the black segment of the population. Early attempts to locate carriers by means of screening programs have met with resistance from some sectors of the black community (*Black Panther*, 1971). Those who object base their argument on the fact that curtailment of marriage and reproduction is essentially genocidal in nature. They believe that attention and research efforts ought to be focused on programs of treatment that do not serve to limit or curtail reproduction. Research is needed to explore the degree to which such resistance prevails in the black community. Screening programs for diseases limited to other segments of the population, such as Tay Sachs disease, which is concentrated in the Ashkenazi Jewish population (Carter, 1969) should be studied. Are these programs meeting similar resistance? Certainly, current social movements and ideologies will have a significant impact on the interpretation and acceptance of such programs by affected segments of the population.

Questions must be asked about the parents who seek genetic counseling. Who are they? Under what conditions do they seek counseling? What types of information do they request? What reactions of guilt or shame do they exhibit about carrying a genetic defect, with the concommitant possibility of passing it on to their children? To what degree do these reactions interfere with effective application of medical genetics? It would be useful to study also the impact of counseling on the social and psychological stability not only of the parents but of the entire family involved. The birth of an abnormal child can be an enormously trying event and can pose many serious adjustment problems for the family.

For example, Birenbaum (1971) has recently reported that the interaction between a mother and a mentally retarded child is significantly different from that between a mother and a mentally normal child. The problems of managing a retarded child are numerous. Birenbaum noted that mothers with mentally abnormal children tended to stress the expressive activities of their role with the child, while with normal children the mothers stressed more the instrumental aspects of the mother-child relationship.

What other problems do abnormal children pose for families, and what is their impact on culturally defined role expectations of parent-child and peer relationships? How do parents respond psychologically to an abnormal child? Research by Olshansky (1962) suggests that a not uncommon response to having a mentally defective child is what he termed "chronic sorrow." This is not necessarily, in Olshansky's view, an unhealthy psychological reaction. Are such responses affected by the type of abnormality a child has? Are responses of families affected by cultural, social, and ethnic variations? A recent investigation by Zuk (1959) suggests that religious beliefs may have a significant impact on the manner in which parents respond to the birth of a defective child. Zuk noted that Catholic mothers more readily accepted a defective child and felt less guilt than non-Catholic mothers. Zuk attributed this to the ability of Catholics to perceive their problem in a meaningful religious context. Zuk's study is highly suggestive, but more extensive research is needed on a wider array of abnormalities and across more religious, ethnic, and social classes.

Not all counselors are concerned with the social and psychological implications of genetic counseling. Studies are needed that reveal the relationship between types of counseling and the adjustment of families to the problems of raising a defective child. Counselors note that there are times when a family is more amenable to counseling than at other times. Parents often have to spend some time adjusting to the birth of an abnormal child before they can consider having another child and can comprehend the significance of risks and other factors brought out in counseling. Research could help to specify the times that are more or less effective and thus contribute to more effective application of medical genetics.

After receiving explanations of Mendelian principles or statements with empirical statistics of risk, parents and families can make a decision concerning future reproductive behavior. We need to know what factors define the range of decisions of those who seek counsel. How important is the physical and social nature of the disease in affecting decisions? What impact do previous births of normal or abnormal children have on future reproductive decisions? The utilization of medical genetics by parents necessarily involves taking risks. At times the risk is minimal, as in the use of amniocentesis to diagnose intra-uterine health (*Fogarty International Center Proceedings*, 1970). At other times the risk can be substantial, as in the case of an autosomal recessive disease such as cystic fibrosis, when parents face a 1 in 4 chance of producing a genetically abnormal child (Carter, 1969). A study by Carter (1966) suggests that in the latter case the degree of risk significantly affects parental decisions to to reproduce or not. However, his data suggest that other factors operate as well. More research of this type is needed to explore the extent to which various social and psychological factors shape reproductive decisions of parents who risk having a defective offspring. Do religious beliefs operate at this stage in the reproductive cycle the way they do after the birth of a defective child (Zuk, 1959)? To what extent does cultural emphasis on the value of child rearing shape parental risk taking, and does this vary culturally?

It is also important to explore parental attitudes toward various forms of genetic abnormality. Is the desire to have perfectly healthy children foremost in parents' minds when they decide about reproductive behavior, adoption, and artificial insemination by a donor? While most parents want the healthiest possible children, some may so desire a child of their own that they will risk disease and its attendant problems. Studies are needed similar to that by Markle and Nam (1971) on desired sex of offspring but focusing instead on parental desires regarding the health of their children. We need to survey couples who decide to adopt and determine the reasons for their choice. What couples decide to use artificial insemination by a donor? What is the impact of artificial insemination by a donor on the social-psychological stability of the family? How does it affect the concept each parent has of himself, and how does it shape parental attitudes toward the child? Existing research suggests that while many couples are willing to use artificial insemination by the husband when necessary, few are willing to use artificial insemination by another donor (Vernon and Boadway, 1959). We need studies that explore the social and psychological factors shaping such attitudes and possible variation in these attitudes across social classes and cultures.

Technical advances in medical science will probably involve various forms of surrogate parenthood, including not only artificial insemination by donors, but possibly egg implants, artificial wombs, and in vitro fertilization (Francoeur, 1970). Existing research suggests that, as in the case of acceptance of birth control, women generally are more receptive to and have more favorable attitudes about such changes in reproduction than do men (Francouer, 1970; Rosenfeld, 1969). While data are limited and must be interpreted carefully, they suggest that most people still consider natural parenthood by the legitimate social parents to be the most desirable state. Use of artificial means of reproduction, except when they benefit the health of the mother, is not favored by any significant segment of the population. More research is needed to increase our understanding of the receptiveness of various populations to changes in reproductive practices.

Legal Considerations

The application of genetic knowledge is complicated by legal constraints. Many problems in medical jurisprudence are related to genetics and run into difficulty for two reasons: sometimes legal precedents do not exist in this area, and the scientific and ethical issues are extremely involved. At present the most prominent problems related to genetics in the field of medical jurisprudence seem to be in these areas: (1) abortion;

(2) consanguineous marriages; (3) marriage between carriers of genetic defects; (4) possible chromosomal basis for criminal behavior; (5) the question of paternal responsibility in cases involving artificial insemination; and (6) malpractice suits involving misdiagnosis of genetic conditions.

Abortion. The development of amniocentesis has brought medical genetics into direct conflict with legal prohibitions on abortion. To a great extent amniocentesis is a useless medical technique unless abortion of a defective fetus is permitted. In fact, many doctors will not perform amniocentesis unless the parents agree to abort the fetus if it is diagnosed as abnormal (Epstein, 1969). This obviously creates a difficult legal situation for the parents as well as the doctor. Because abortion of a defective fetus is presently permitted in approximately fourteen states, the use of this form of medical genetics is limited (Lader, 1970). Nevertheless, amniocentesis is being used more and more, and research is needed on the ways in which doctors and patients contend with the legal aspects of aborting a defective fetus.

Although considerable research has been conducted on abortion, a major focus has been on abortion of an unwanted pregnancy, not on abortion of a potentially abnormal child (Lee, 1969; Rosen, 1967). We need studies that outline both professional and public responses to therapeutic abortion of fetuses with varying degrees of abnormality. Packer and Gampell (1959) surveyed doctors on their attitudes and behavior concerning abortions performed for various reasons, including the therapeutic abortion of a defective fetus. The authors reported that while 84 percent of the doctors interviewed would perform an abortion if the health of the mother was threatened, only 19 percent would abort a fetus if it was abnormal. Because intra-uterine diagnosis has been improved since their investigation, their study is dated and should be conducted again.

Even in states that permit the abortion of a defective fetus, the final decision regarding particular cases is often the responsibility of a hospital committee. Many hospitals currently require that the abortion of a fetus be performed only with the consent of a group of doctors who must weigh the evidence and decide if an abortion would be reasonable. Since physicians' cumulative knowledge and attitudes about what constitutes a serious risk thus often determine the use of abortion, studies are needed concerning the attitudes of doctors toward genetic disease and statements of risk. Sometimes doctors overrule the desires of a particular couple by not permitting an abortion and letting the fetus carry to term. If in such a case the fetus is defective, are the doctors legally responsible for permitting such a child to be born?

CONSANGUINEOUS MARRIAGES. Legal concern with consanguineous marriages dates back to ancient history, when religious precepts forbade incest and the marriage of relatives. Legal concern in the United States dates back to the late nineteenth and early twentieth centuries. A recent survey of the laws in the United States by Farrow and Juberg (1969) reveals considerable variability from state to state. Most states prohibit marriage between persons with a coefficient of relatedness equivalent to first cousins or closer. The rationale underlying these laws varies, sometimes being based on religious beliefs, sometimes on the deleterious genetic consequences of offspring born to related individuals.

MARRIAGE BETWEEN CARRIERS OF GENETIC DEFECTS. Although recent developments in medical genetics permit the detection of an increasing number of carriers of genetic defects, there are no legal constraints on marriage between carriers. The same legal rationale that prohibits marriage between relatives and mental defectives might be used to prohibit marriage between carriers. The legal implications of developments in carrier detection and screening programs must be explored. No states in the United States have instituted mass programs to screen genetic disease, except for phenylketonuria (PKU), a metabolic disturbance resulting in severe mental retardation unless diagnosed early and treated by dietary control. Screening programs for PKU have recently come under attack because of technical problems in diagnosing the disorder and the consequences of the treatment program (Danes, 1970). Exploration of the conditions leading to the early and extensive use of PKU screening might shed light on the acceptance of further mass screening programs. The legal implications of screening programs need study, with special attention given to their potential impact on marriage and divorce laws.

THE XYY ANOMALY. Recent controversy over the correlation between a specific chromosomal defect, the XYY abnormality, and various forms of pathological social behavior brings to the forefront an important issue in applied human genetics (Epstein, 1969). If social behavior may have a significant genetic component, a new concept of legal responsibility is introduced.

Numerous articles have appeared outlining the nature of the chromosomal abnormality and its possible relationship to deviant behavior (Epstein, 1969). The best evidence to date suggests that no clear relationship has been established between this chromosomal complement and social pathological behavior (Public Health Service, 1970). Since the XYY chromosomal abnormality can be detected in utero by amniocentesis, the legal problems discussed above under "abortion" arise again.

ARTIFICIAL INSEMINATION. Adoption used to be the only form of surrogate parenthood, but, today, various forms of surrogate parenthood are becoming possible through advances in medical genetics. Through artificial insemination by a donor, it is possible for a woman to be married to one man and legally mother a child by another. Artificial insemination is used not only when the husband is sterile, but also when the husband carries a defect, or even when both parents are carriers of an anomaly and their children would run a high risk of abnormality. The practice of artificial insemination has already given rise to considerable legal controversy (Epstein, 1969). The rights and duties of the social and biological fathers are not yet clearly defined legally. Research is needed on the legal problems involved in such situations. What are the possible implications for inheritance laws? In addition to the social and psychological impact on the family, does artificial insemination pose any legal problems different from those posed by adoption? Do parents have a right to know the biological father?

PROBLEMS OF MISDIAGNOSIS. Finally, the problem of misdiagnosis has always constituted a major legal issue for the medical profession, and this is no less true with medical genetics. Today, because most physicians lack adequate training in diagnosing genetic defects, misdiagnosis and inappropriate advice can be serious problems. For example, if a couple are told that a given problem is genetic and accordingly opt for sterilization, they have taken an irreversible step. If the doctor was wrong, not only has he caused the couple much grief, but he is legally liable. The current structure of genetic counseling facilities combined with the lack of diagnostic capacity make such problems likely. In addition, given the current lack of training in medical genetics, practicing physicians probably ignore the genetic aspects of many diseases. In such situations what legal responsibility does the physician have to inform his patients of the genetic basis of their disease? We need studies of such problems and the impact they have on rules guiding counselors' and general practitioners' decisions in their daily use of medical genetics.

Economic Aspects of Applied Genetics

As Bailey (1970) recently noted, the medical profession has traditionally been reluctant to interpret its work in terms of economic considerations. Nevertheless, economic constraints always shape the delivery and utilization patterns of medical practice, if not in a primary at least in important secondary ways. The impact of economic factors on medical genetics can be explored on two levels: (1) the family level, with an analysis of the costs of a genetically defective child, and (2) the societal level, with an analysis of the costs to local, state, and federal agencies of maintenance of a segment of its population with genetic problems.

COSTS TO THE FAMILY. The birth of a genetically defective child can pose many problems of resource allocation for a family. The maintenance of a child with a severe anomaly can cost a family thousands of dollars a year. The assignment of family economic resources to such a problem can often leave few resources for more normal and general family expenses. How do families allocate their financial resources when faced with such a problem? What are the sources of support? Do patterns of family financial assistance in this situation vary by social class, religious affiliation, and ethnic group? It would seem that exposure to the economic burden of a handicapped child could have a significant impact on the decisions of parents to risk having additional handicapped children. To what extent does the economic burden of a genetically defective child shape parental decisions about future reproductive behavior?

Genetic counseling can itself be expensive, if extensive laboratory work is involved and repeated use is made of amniocentesis or other methods of intra-uterine detection. Studies are needed of the costs involved in the delivery of genetic counseling to various segments of the population. Also, it is important to analyze the degree to which the financial condition of a family prevents it from seeking medical genetic services. While the economic costs of genetic counseling have probably been relatively unimportant in shaping utilization patterns in the past, this is changing with the increasing inclusion of clinical genetics into medical practice.

COSTS TO SOCIETY. Estimating the costs of genetic defects to society is enormously complex. A few attempts have been made to estimate the costs and economic benefits of various types of screening programs. Otherwise, little work has been done in the area and the literature is very limited.

The costs to society of genetically defective persons are more than medical. For example, society has to aid not only in the support and maintenance of these individuals, but it has to forego their potential productivity and contributions, economic and social. Genetic diseases constitute a significant proportion of institutional cases. A recent study by Stevenson (1959) reports that approximately 26 percent of all institutional beds in Northern Ireland were occupied by individuals with genetically determined illnesses. Since most genetic disorders are chronic in nature, it is apparent that institutionalization can become very extensive and expensive.

Wright's (1960) theoretical approach to the problem of estimating the many costs to society of genetically determined diseases is an important contribution to economic analysis in this area. He considers various aspects of the disease, such as time of onset, duration, and the physical and mental nature of the defect, as determinants of costs to society. More such studies outlining the costs of various types of genetic diseases and various types of institutionalization will be basic to the formulation of genetic programs at any level of government and to the implementation of mass medical genetic policies.

Critical economic analysis of genetic problems will have to include cost-benefit analysis of preventive as opposed to treatment programs. While treatment is available for many genetic defects, they could often be avoided altogether through monitoring of pregnancies with high risk of defects. Even if such analysis were to reveal that prevention is considerably more economical than treatment, the many social and religious objections to the major form of prevention today, abortion, would impede the implementation of genetic policies simply on the basis of their economic feasibility.

A few studies attempt to estimate directly the medical costs involved in the treatment of specific genetic diseases and the costs involved in various preventive programs, such as mass screening and intra-uterine diagnosis (Scriver, 1970; Danes, 1970). Some states have instituted mass screening programs for phenylketonuria, and comparisons have been made between the costs of these programs and the costs to the state if the children with PKU had not been screened and treated. For example, in Massachusetts it costs the state about $175,000 a year to screen approximately 97 percent of all newborns for PKU. In an average year about eight children are identified as having the disease. If these children were not identified and treated, the cost to the state would be approximately $150,000 to $200,000 per lifetime for each child (Danes, 1970). While this suggests an impressive cost-benefit ratio for PKU screening, additional studies of this type are needed before we will have reliable estimates of the direct cost to society of genetic anomalies and can make wise decisions concerning prevention and treatment programs.

It is also important that we explore more fully such topics as measuring an individual's value to society, the extent to which society should bear the costs of genetically defective individuals (especially when the society does not permit abortion of defective fetuses), and the extent to which medical genetics is and can be covered by various forms of medical insurance, public and private. Discussion of these issues is needed, in the fashion of Bailey's (1970) recent analysis of the economic and social costs of death.

It can be argued that the economic problems and issues involved in the application of genetic knowledge to human populations are not as important as the social, psychological, and moral issues. However, economic constraints often shape and dictate important decisions concerning the use of medical knowledge. The studies outlined above will provide needed information on present costs and probable payoffs of enlarged medical genetic programs. Such information must be included in any discussion of medical genetics and must be used in arriving at decisions about the use of medical genetics for the individual's and society's benefit.

References

Babbie, Earl R., *Science and Morality in Medicine*. Berkeley: University of California Press, 1970.

Bailey, Richard M., "Economic and Social Costs of Death," 275-302 in O. B. Brim, H. E. Freeman, S. Levine, and N. A. Scotch, eds., *The Dying Patient*. New York: Russell Sage Foundation, 1970.

Birenbaum, Arnold, "The Mentally Retarded Child in the Home and Family Cycle," *Journal of Health and Social Behavior,* 12 (March), 1971, 55-65.

Black Panther, 6, No. 2 (May 22, 1971) and 6, No. 17 (April 10, 1971).

Bloom, Samuel W., *The Doctor and His Patient*. New York: Russell Sage Foundation, 1963.

Carter, Cedric, "Comments on Genetic Counseling," Proceedings of the Third International Congress of Human Genetics, 1966, 97-100.

Carter, Cedric, *An ABC of Medical Genetics*. Boston: Little, Brown, 1969.

Clausen, John A., Morton A. Seidenfeld, and Leila C. Deasy, "Parent Attitude toward Participation of their Children in Polio Vaccine Trials," *American Journal of Public Health* 44 (1954), 1526-1536.

Crane, Diana, *Social Aspects of the Prolongation of Life,* 1, Social Science Frontiers. New York: Russell Sage Foundation, 1969.

Danes, Betty Shannon, "Genetic Counseling," *Medical World News* (November 6, 1970), 35-41.

Deasy, Leila, "Socio-Economic Status and Participation in the Poliomyelitis Vaccine Trials," *American Sociological Review,* 24 (1956), 185-191.

Deshaies, J., "Public Knowledge about Chronic Disease Symptoms," unpublished masters thesis, University of Chicago (1962), cited in Feldman (1966).

Epstein, Charles J., "Medical Genetics: Recent Advances with Legal Implications," *Hastings Law Journal,* 21, No. 1 (November, 1969), 35-49.

Farrow, Michael, and Richard Juberg, "Genetics and Laws 'Prohibiting' Marriage in the United States," *Journal of the A.M.A.,* 209 (July 28, 1969), 534-538.

Feldman, Jacob J., *The Dissemination of Health Information.* Chicago: Aldine, 1966.

Fletcher, John "Parents in Genetic Counseling: The Moral Shape of Decision Making." Paper presented at Conference on Ethical Issues in Genetic Counseling and the Use of Genetic Knowledge, Fogarty International Center, National Institutes of Health, October, 1971.

Fogarty International Center Proceedings, No. 6 (May, 1970), Maureen Harris, ed.

Francoeur, Robert T., *Utopian Motherhood.* Garden City, N.Y.: Doubleday, 1970.

Freidson, Eliot, *Profession of Medicine.* New York: Dodd, Mead, 1970.

Juberg, Richard, "Heredity Counseling," *Nurses Outlook,* 14 (January, 1966), 28-33.

Koos, E. L., *The Health of Regionville.* New York: Columbia University Press, 1954.

Lader, Lawrence, "A National Guide to Legal Abortion," *Ladies Home Journal* (July, 1970), 73.

Lee, Nancy Howell, *The Search for an Abortionist.* Chicago: University of Chicago Press, 1969.

Lindzey, Gardner, John Loehlin, Martin Manosevitz, and Delbert Thiessen, "Behavioral Genetics," *Annual Review of Psychology,* 1971.

Lynch, H. T. *Dynamic Genetic Counseling for Clinicians.* Springfield, Ill.; Charles C. Thomas, 1969.

Lynch, H. T., and D. Bergsma, *Birth Defects Genetic Services.* New York: National Foundation–March of Dimes, 1971.

Lynch, H. T., Robert L. Tips, Anne Krush, and Charles Magnuson, "Family-Centered Genetic Counseling: Role of the Physician and Medical Genetics Clinic," *Nebraska State Medical Journal,* 5, No. 4 (April, 1965), 155-159.

Markle, Gerald E., and Charles B. Nam, "Sex Predetermination: Its Impact on Fertility," *Social Biology,* 18, No. 1 (March, 1971), 73-83.

Mechanic, D., *Medical Sociology: A Selective View.* New York: Free Press, 1968.

Menzel, Herbert, and Elihu Katz, "Social Relations and Innovations in the Medical Profession: The Epidemiology of a New Drug," *Public Opinion Quarterly,* 19 (Winter, 1955-1956), 337-352.

Miller, James R., "Human Genetics in Public Health Research and Programming," *Symposium on Human Genetics in Public Health,* Minneapolis (August 9, 1964).

Myers, Jerome K., and L. Schaffer, "Social Stratification in Psychiatric Practice: A Study of an Outpatient Clinic," *American Sociological Review,* 19 (June, 1954), 307-310.

Neel, James V., "The Meaning of Empiric Risk Figures for Disease or Defect," *Eugenics Quarterly,* 5 (March, 1958), 41-43.

Olshansky, Samuel, "Chronic Sorrow—A Response to Having a Mentally Defective Child," *Social Casework,* 43, No. 4 (April, 1962), 190-193.

Packer, H. L. and R. J. Gampell, "Therapeutic Abortion: A Problem in Law and Medicine, "*Stanford Law Review,* 11 (1959), 417.

Public Health Service, "Report on the XYY Chromosome Abnormality," *Public Health Service Publication,* No. 2103 (October, 1970).

Reed, S. C., *Counseling in Medical Genetics,* 2nd ed. Philadelphia: Sanders, 1963.

Reismann, Leonard E. and Adam P. Matheney, *Genetics and Counseling in Medical Practice.* St. Louis: C. V. Mosby Company, 1969.

Rosen: Harold, *Abortion in America.* Boston: Beacon Press, 1967.

Rosenfeld, Albert, *The Second Genesis: The Coming Control of Life.* Englewood Cliffs: Prentice Hall, 1969.

Scriver, Charles R., "Screening for Inherited Traits: Perspectives," *Fogarty International Center Proceedings,* No. 6 (May, 1970) Maureen Harris, ed.

Sorenson, James R., "Genetic Counseling Bibliography." Unpublished manuscript, Princeton University, 1971a.

Sorenson, James R. "Decision Making in Applied Human Genetics: Individual and Societal Perspectives." Paper presented at Conference on Ethical Issues in Genetic Counseling and the Use of Genetic Knowledge, Fogarty International Center, N. I. H., October, 1971b.

Stevenson, A. C., "The Load of Hereditary Defects in Human Populations," *Radiation Research,* Supplement 1 (1959), 306-325.

Symposium on Human Genetics in Public Health. Minneapolis (August 9-11, 1964).

Szasz, T. S., and M. H. Hollender, "A Contribution to the Philosophy of Medicine: The Basic Models of the Doctor-Patient Relationship," *A.M.A. Archives of Internal Medicine,* 97 (May, 1956), 585.

Vernon, Glenn M., and Jack A. Boadway, "Attitudes toward Artificial Insemination and Some Variables Associated Therewith," *Marriage and Family Living,* 21 (1959), 43-47.

Wilson, Robert N., "The Physician's Hospital Role," 414-415 in *Medical Care: Readings in the Sociology of Medical Institutions.* W. Richard Scott and Edmund H. Volkart, eds., New York: Wiley, 1966.

World Health Organization, *The Teaching of Genetics in the Undergraduate Medical Curriculum and in Postgraduate Medicine,* Technical Report Series, No. 238 (1962).

Wright, Sewell, *The Biological Effects of Atomic Radition.* Washington, D. C.: National Acaedmy of Sciences–National Research Council. 1960.

Zuk, G. H., "The Religion Factor and the Role of Guilt in Parental Acceptance of the Retarded Child," *American Journal of Mental Deficiency,* 64, No. 1 (July, 1959), 139-147.